I AM EDDE EQUAL

By Diana T. Phillips

Illustrations By K. K. P. Dananjali

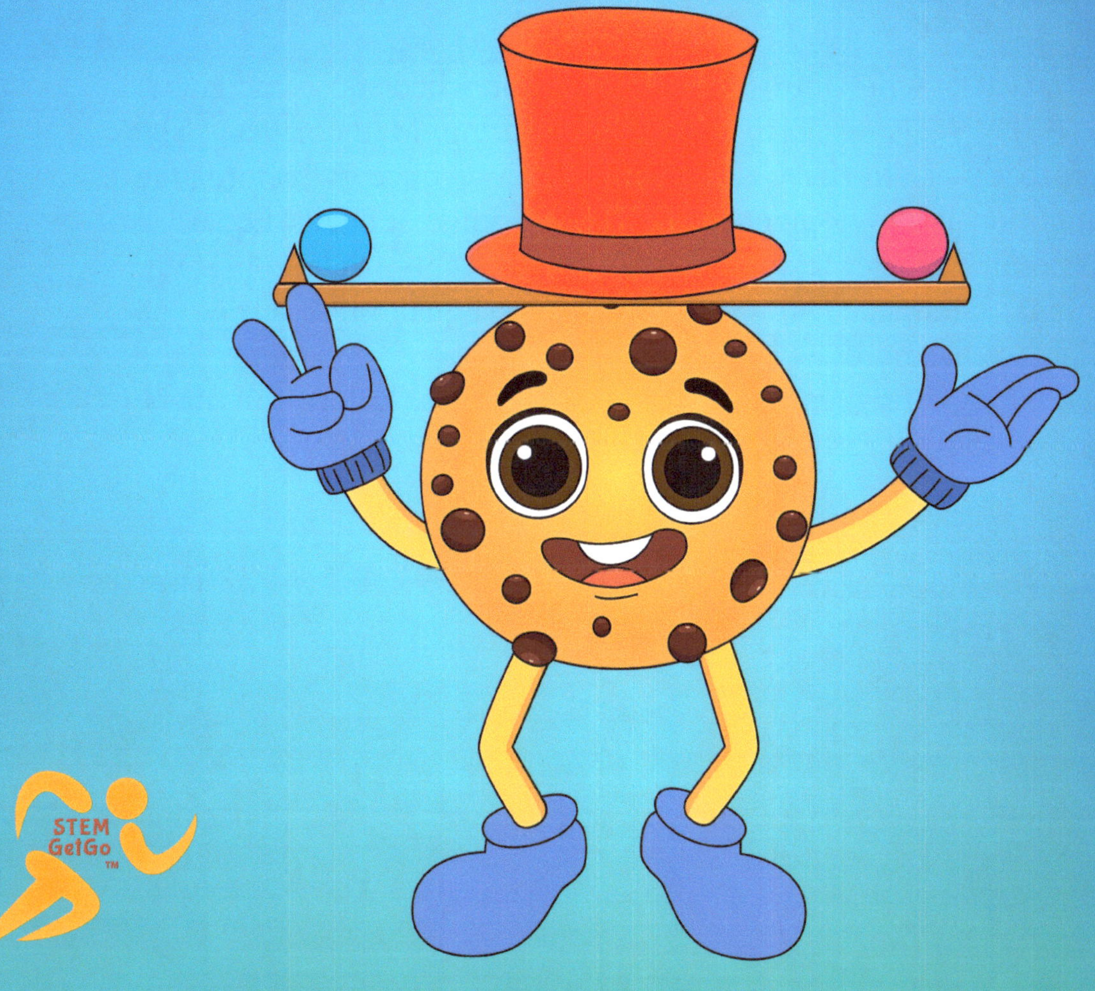

StemGetGo series introduces children (preschoolers – 2nd grade) to STEM (Science, Technology, Engineering, Mathematics). The approach is to have fun with STEM concepts before the intentional development of STEM skills. EDDE Equal, a cartoon character, experiences STEM concepts while doing fun things. Students will see that EDDE Equal must always be equal or balanced, a fundamental concept of equations. The series uses fundamental skills (sight words, colors, numbers, and shapes) of early childhood education.

HELLO!
I AM EDDE EQUAL

By Diana T. Phillips

Illustrations By K. K. P. Dananjali

To my little heartbeats who helped decide what EDDE Equal should look like:

Addyson, Jordyn, Kaleb, Mikyle, Waylin, Winifred, and Wylder.

To Marsha Sherrod and Mona Richardson for unwavering support.

To Grady Jessup, Esq., for numerous reviews, invaluable feedback, and support.

To Kyle for love and support.

P.O. Box 80595
Raleigh, NC 27623

ISBN: 979-8-9942224-0-9

Hello! I am EDDE EQUAL.
My first name is EDDE.
My last name is EQUAL.

I look like a cookie.
I wear a bright red hat.
I wear shiny blue gloves.
I have a purple cat.

I love my bright red hat.
It's always on my head.
I only take it off
Before I go to bed.

There's a teeter-totter line
That's under my red hat.
The line looks very funny
Just like my purple cat.

My teeter-totter line can move.
It moves up and down.
My hat must stay in the center.
It can't touch the ground.

Awful things will happen
When my hat falls to the ground.
I get very frightened,
My heart starts to pound and pound.

My teeter-totter line holds balls.
Balls are on the left of my hat
Balls are on the right.
All are the same size.
And some glow at night.

With balls on the left and right
I play a fun game.
The name of the fun game is
KEEP THE BALLS THE SAME.

What exactly do I mean?
Do I keep colors the same?
No, that's not what I mean at all.
I don't care about colors of balls.

Colors can be different.
Colors can be the same.
Always remember
The name of this game.

The name of this game is
KEEP THE BALLS THE SAME.
But what about the balls?
WHAT MUST BE THE SAME?

Come along with me.
I'll give you a clue.
The clue will tell you
What you have to do.

I add 1 ball on the left.
I add 1 ball on the right.
I'm standing very tall.
My 2 balls will not fall.
Right
Left

I add 2 balls on the left.
I add 2 balls on the right.
I'm standing very tall.
My 4 balls will not fall.

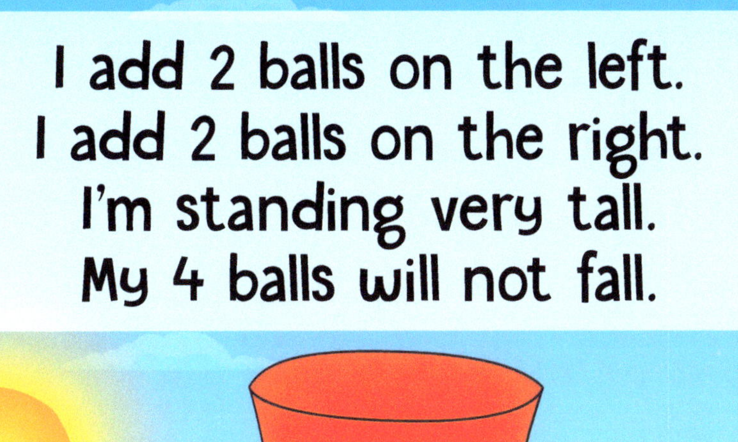

Right

Left

I put 1 ball on the left.
I put 3 balls on the right.
Oh my! My hat is falling down!
And I am falling to the ground!

I must act quickly!
I know what to do!
I need more balls!
I need only 2!

I will add 2 more balls
Where there's only 1.
Then I will be EQUAL,
And I can have fun!

EDDE EQUAL is my name.
The number of balls
on each side of my red hat
MUST BE THE SAME.

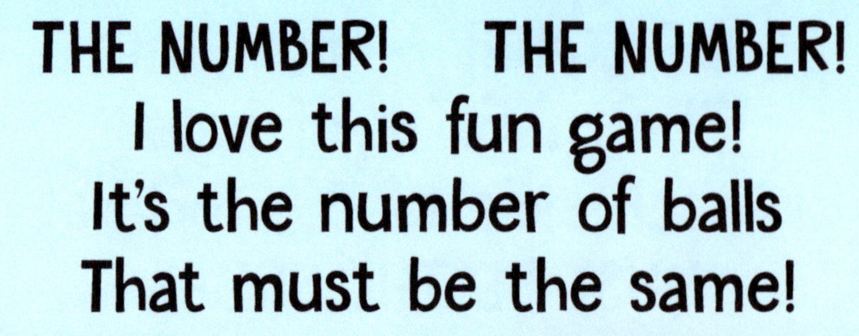
THE NUMBER! THE NUMBER!
I love this fun game!
It's the number of balls
That must be the same!

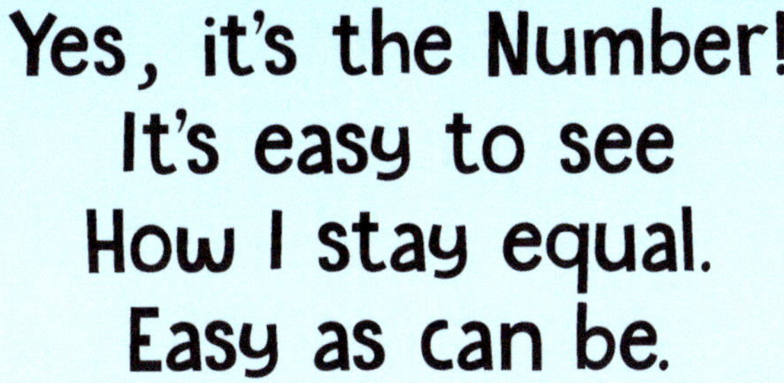
Yes, it's the Number!
It's easy to see
How I stay equal.
Easy as can be.

Now I can run!
I can have fun!
I can wear my shiny red hat!
I can play with my purple cat!

EDDE EQUAL is my name.
Balls on both sides of my hat
HAS TO ALWAYS BE THE SAME.

I love my name.
It tells you about me.
When I'm equal,
I'm happy as can be.

My first name is EDDE.
My last name is EQUAL.

SIGHT WORDS TO LEARN/REVIEW

- I
- my
- like
- to
- go
- and
- in
- a
- are
- thing/things
- then
- what
- read
- ball
- hat
- cat
- run
- fun

NEW WORDS TO LEARN

- equal
- same
- different
- number
- wear
- purple
- down
- keep
- clue

SKILLS INTRODUCED / LEARNED

Fundamental of Equations

Equal / Not Equal

www.ingramcontent.com/pod-product-compliance
Lightning Source LLC
Chambersburg PA
CBHW042026110726
48010CB00007B/241
9798994222409